Roadside Geology along Sunrise Highway

Photo by David M. Bloom

I0809467

Scene at Mile 18.8. Photograph by Philip T. Farquharson

Roadside Geology along Sunrise Highway

San Diego County, California

Dr. Michael J. Walawender

Emeritus Professor

Department of Geological Sciences

San Diego State University

SAN DIEGO ASSOCIATION OF GEOLOGISTS

www.sandiegogeologists.org

Distributed by

Sunbelt Publications

www.sunbeltpublications.com

Roadside Geology along Sunrise Highway, San Diego County, California

 First edition 2012, seventh printing 2025.

Edited by Carole L. Ziegler, SDAGeditor@gmail.com
Cover and book design by Philip T. Farquharson, geoguy@cg-squared.com

Please direct comments and inquiries to:
San Diego Association of Geologists
3130 North Evergreen Street
San Diego, California 92110
www.sandiegogeologists.org

Available from:
Sunbelt Publications, Inc.
664 Marsat Court, Suite A, Chula Vista, CA 91911
(619) 258-4911, fax: (619) 258-4916
www.sunbeltpublications.com
mail@sunbeltpub.com

28 27 26 25 10 9 8 7

Cataloging-in-Publication Data

Walawender, Michael J., 1940-2011
Roadside Geology along Sunrise Highway / Michael J. Walawender
p. cm.
Includes bibliographic references, glossary and mineralogy index
ISBN 978-0-916251-19-2
1. Geology—California—San Diego County Highway S1. 2. Guidebooks and travel—California—San Diego County. 3. Geology—California—San Diego—Laguna Mountains. I. Walawender, M.J. II .Ziegler, Carole, L. III. Farquharson, Philip T.

QE90.S 2012
557.94'98—dc23

Cover photograph courtesy of Philip T. Farquharson, View from Mile 37 of Sunrise Highway across Cuyamaca Meadows and Lake to Stonewall Peak.
Front inside — General Geology of San Diego County and Field Trip Route, Michael J. Walawender.
Back inside — Field Trip Stops and Geologic Map of a Portion of the Cuyamaca-Laguna Mountain Shear Zone, modified by David M. Bloom.
All photos and figures by Michael J. Walawender, unless otherwise credited.

Dedication

For Sonja and Joshua Walawender

Table of Contents

List of Figures

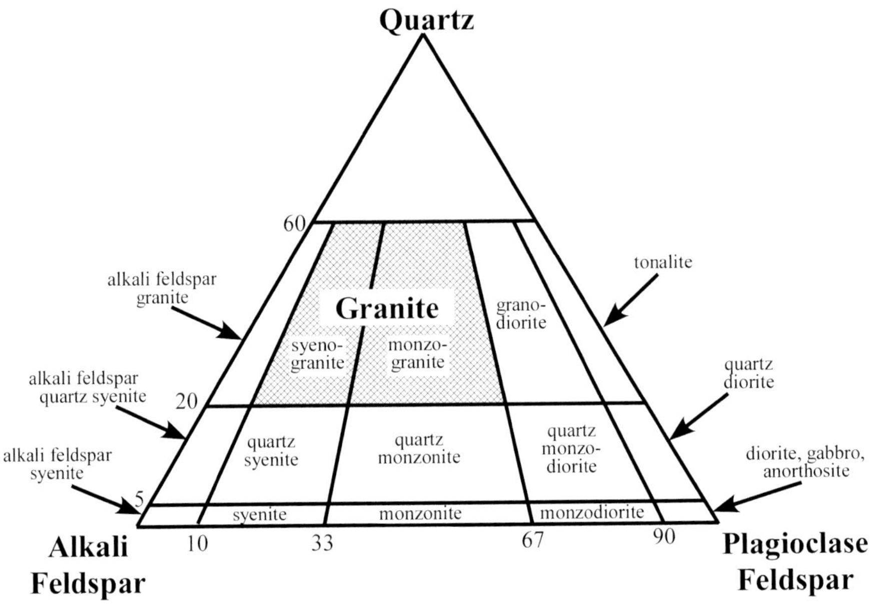

Figure 1. Classification Scheme for Granitic Rocks. The rock name is based on the relative percentages of the three framework minerals, quartz, plagioclase feldspar, and alkali feldspar. Based on Streckeisen (1974).

Introduction

Sunrise Highway (San Diego County Highway S1) is designated as a National Forest Scenic Byway and follows the backbone of the Laguna Mountains northward towards the historic gold mining community of Julian. This route approximates an important geologic boundary that divides the Mesozoic Peninsular Ranges Batholith into eastern and western zones that have their own distinctive features (inside front cover). The Cretaceous age plutonic rocks in both zones intrude older metasedimentary rocks of probable Triassic age. Recently, a third component has been recognized whose complexities add a new dimension to the geologic history of the region. Scattered bodies of deformed Jurassic granitic rocks occur along a discontinuous belt within the batholith's western zone and will be the backdrop for much of this drive (Figure 2).

Recent paleomagnetic studies suggest that the entire batholith had a very dynamic and kinematic history. Although the eastern zone appears to have essentially formed in place, the western assemblage represents an older exotic terrane that has drifted several hundred kilometers northwestward to its present location. Its collision with Jurassic rocks near the western edge of the North American continent at approximately 110 million years ago created the complex mix of ages, structures, and lithologies that you will see on this drive.

Western Zone Plutonic Rocks

Three main plutonic rock types, gabbro, monzogranite, and tonalite (Figure 1) occur in the western zone of the batholith. They form modest-sized [less than 40 square miles (100 km^2)] plutonic bodies that can be recognized by their distinctive morphology. The earliest magmatic activity in this zone of the batholith appears to be represented by rocks of gabbroic composition. Radiometric ages on these rocks, however, have been difficult to obtain due to their low concentrations of zircon or other dateable materials. They have been eroded into topographic prominence to form landmarks such as Cuyamaca Peak, Viejas Mountain, and Los Piños Mountain that are characterized by sparse, dark-colored outcrops with deep brown to reddish soil cover.

The monzogranite plutons also form prominent ridges and peaks but are characterized by light-colored, bouldery outcrops. They underlie topographic landmarks such as Woodson Mountain near Ramona and Chiquito Peak adjacent to Interstate Highway 8 just east of the Viejas Indian Reservation. Emplacement ages within this rock type fall between 115 and 122 million years ago.

The tonalitic plutons underlie relatively low, rolling terranes with scattered, dark gray outcrops and grayish soils. Locally, this rock type contains prominent, dark, elongate fragments of older igneous rocks up to two feet in length that were caught up in the rising magma. Interstate Highway 8 lies within the elongate Alpine (tonalite) Pluton from Flinn Springs to the Pine Valley Creek Bridge. Radiometric ages typically fall between 102 and 110 million years ago.

Paleomagnetic studies on the tonalitic and gabbroic rocks indicate that they (and the whole of the western zone) formed up to 250 miles (400 kms) to the south-southwest and must have migrated to their present position where they collided with a Jurassic arc outboard of the western edge of the North American continent approximately 110 million years ago.

The different topographic expressions of the three main rock types within the western zone were enhanced during the Miocene Epoch when the climate was much wetter than today. Warmer temperatures, elevated concentrations of carbon dioxide, and abundant decaying vegetation combined with rainfall to chemically weather the near-surface rocks. The degree of chemical weathering was, in good measure, controlled by the variation in fracturing in the three rock systems. The gabbro and monzogranite plutons appear to be less fractured and to have undergone less intense chemical weathering than the tonalites so that erosion by running water would preferentially remove the weathered tonalite debris leaving the other two rocks types as topographic highs.

Acknowledgements

Michael "Mike" Walawender wanted to leave one last legacy to the San Diego community and its visitors. As Mike was first developing this guidebook, he was accompanied by his wife, Sonja, to decide upon stops and work out the mileage. Then came many more drives along this beautiful highway by Mike Hart, Greg Cranham, Dave Bloom, Monte Marshall, Monte and Diane Murbach, and myself to make sure that the self-guided tour could be easily followed by both the professional geologist and the casual visitor. The record number of reconnaissance trips, however, goes to Phil Farquharson who made the trip at least 5 times to double check mileage, GPS waypoints, and hiking miles. David Bloom and Phil Farquharson worked their usual magic on figures and maps and generously donated their photographs to supplement Mike's. An extra-special thank you goes to Mary Walke, who opened up her cabin in the Lagunas for a little rest, libation and camaraderie on these forays. Monte Murbach has gained a new nickname, "Sherpa", for the many duties that he volunteered to take on as Assistant Managing Editor for this project.

This book would not have seen the light of day if it weren't for the support of the many members of the San Diego Association of Geologists, both in their encouragement and in the funding of this project.

Special recognition goes to the San Diego State Geology Alumni Association. They have adopted the northern two miles of Sunrise Highway as a volunteer project and every three months, they descend upon the highway with their orange bags, gloves and grabber sticks to make sure that the highway remains as beautiful as it is. Thank you Mike Hart for managing this project.

Thank you rain, for watering the wildflowers so that we may all enjoy them as much as Mike did.

Carole L. Ziegler

Managing Editor

San Diego Association of Geologists

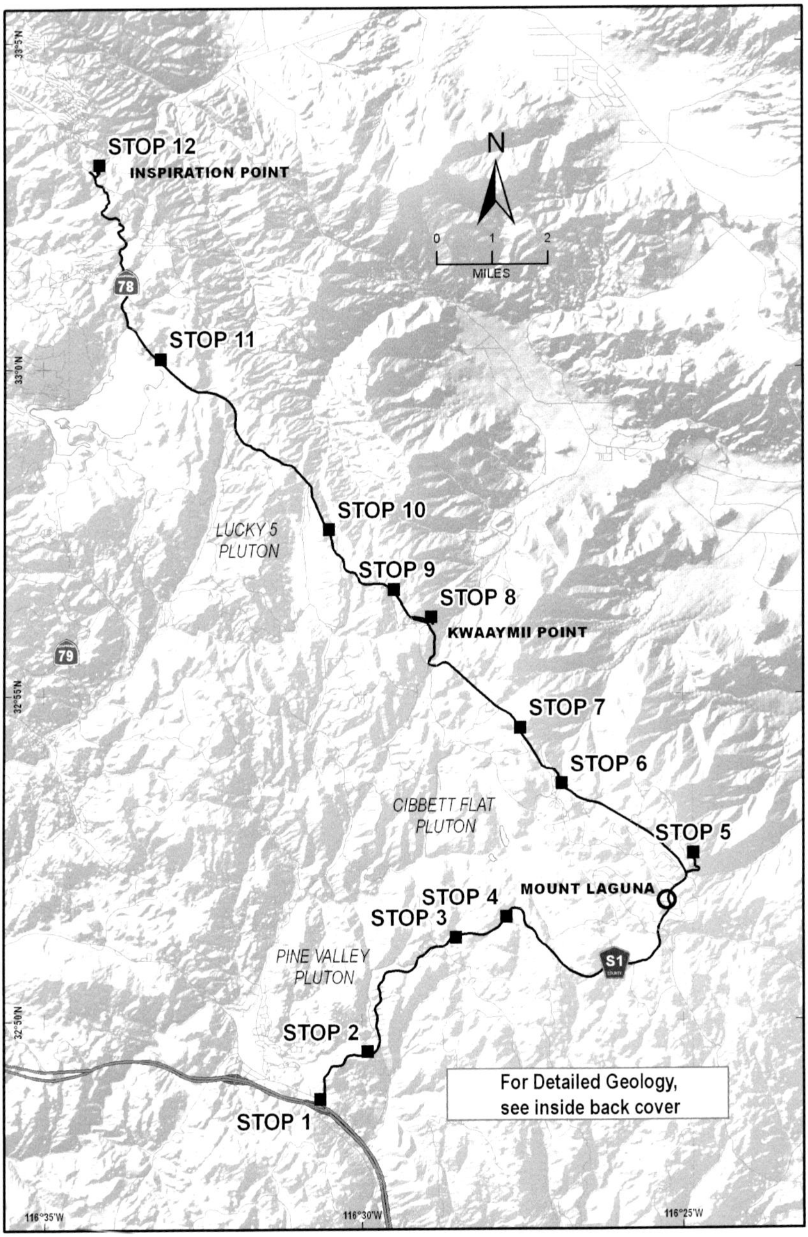

Figure 2. Field Trip Stops of a Portion of the Cuyamaca-Laguna Mountain Shear Zone. After Todd (2004), adapted by David M. Bloom. (for detailed geology in full color, see inside back cover).

ROAD LOG

Intersection of Old Highway 80 and Sunrise Highway (County Highway S1), 47 miles from the city of San Diego, Exit 47 on Highway 8. *Mile Posts are small green signs found along the eastern side of the highway. Numbers in bold on the left side of the text represents either one of these mile posts or an extrapolation to tenths of a mile based on these mile posts at stops or points of interest along Sunrise Highway. Numbers in italics indicate the elapsed mileage from the start (not including any side-trips).*

Please note: Either a Forest Adventure Pass or National Parks and Federal Recreational Lands Pass is required to park along Sunrise Highway. Forest Adventure Passes are available at the Pine Valley Store, Mountain Market, Blue Jay Lodge, the Laguna Mountain Visitor Center, Laguna Mountain Lodge, Big 5 Stores, REI and Adventure 16. The National Parks and Federal Recreational Lands Passes (formerly known as Golden Eagle or Golden Age Passports) are available by mail or from National Park Service units, such as Cabrillo National Monument or Joshua Tree National Park.

13.8 *0.0* **STOP 1. Really Weird Rocks**. [32.81311°N, 116.50908°W] [elevation 4,111 feet] Park along the wide shoulder on the east side of Sunrise Highway (S1) just north of the Interstate Highway 8 overpass. Walk about one hundred yards along the south side of Old Highway 80 to view outcrops of the 118 million year old Pine Valley Monzogranite and the unusual rocks intruded into it.

Weird Rock 1. [Mile Post 6.5 on Old Highway 80] [32.81328°N, 116.50996°W] [elevation 4,101 feet] A few yards west of the Sunrise Highway intersection along Old Highway 80, there is a zone of pegmatitic granite surrounding massive milky to gray quartz with large blocks (crystals?) of alkali feldspar. This zone has sharp contacts with and appears to be intrusive into the Pine Valley Monzogranite. To the west and across the road in the brush, other large outcrops of similar material are visible. The abandoned gold mines in Noble Canyon just a mile or so to the northwest followed one such vertical system.

These rocks may represent a vertical zonation in the gem-bearing pegmatites com-mon in the medial portions of the batholith. In such a view, the gem-bearing portion of the pegmatites give way upward into barren quartz and feldspar, and, at still shallower (but not necessarily shallow) depths, pass into gold-bearing quartz dikes comparable to those found near Julian, California. Similar changes have been described in the Bolivian Andes where granitic pegmatites pass upwards into cassiterite-quartz veins and eventually into barren quartz dikes.

Weird Rock 2. [32.81395°N, 116.51144°W] [elevation 4,069 feet] Continue west down the shoulder another few hundred feet to view a narrow dike (?) of essentially pure biotite cutting the Pine Valley Monzogranite (Figure 3). This rock, by virtue of its coarse grain size and lack of a prominent foliation direction appears to be magmatic in origin but how such rocks form is a matter of conjecture. The only name that comes to mind for this

Figure 3. Irregular Body of Biotite Lamprophyre Encased in Pine Valley Monzogranite.

rock is the enigmatic term "lamprophyre". Biotite lamprophyre is commonly associated with calc-alkaline igneous rocks and may contain subordinate amounts of amphibole, various pyroxenes, feldspar, and olivine. Megascopically, however, this rock appears to be devoid of all these trace minerals. What is a lamprophyre, you ask? No one seems to know. Consider this quote from John Winter's petrology text. "During most of my petrological career, I never understood lamprophyres well at all. Through research and discussions, I discovered that I am in good and plentiful company." (Winter, 2001)

PLEASE DO NOT TAKE SAMPLES. IT'S A VERY SMALL OUTCROP AND SHOULD BE PRESERVED FOR OTHERS TO VIEW AND CONSIDER!

Return to your car confused but vowing to look up the term "lamprophyre" somewhere. Head north on Sunrise Highway.

14.0 Reddish outcrops on the right (northeast) are weathered olivine-rich gabbroic rocks. This is the southern end of a small pluton that is composed of two distinct gabbroic suites. This southern side mainly contains olivine-rich rocks that vary from peridotite to olivine gabbro to minor anorthosite. Olivine is a magnesium and iron-bearing silicate mineral which, when exposed to the atmosphere, undergoes oxidation. The reddish color is due to oxidation of the iron in olivine which converts it from Fe^{++} (ferrous) to Fe^{+++} (ferric). The Fe^{+++} is insoluble in surface waters and remains behind to stain the rock. The northern half of this pluton consists of hornblende-rich gabbro and lacks the olivine and reddish staining found in the southern half.

14.7 Saddle with parking area to right (east). This saddle marks the approximate boundary between the two halves of the gabbro pluton. Exploratory drilling at this spot several years ago brought up more than 300 feet of hornblende-rich gabbro.

15.3 Road cut exposes a complex fault array juxtaposing weathered buff-colored Jurassic gneiss against Cretaceous hornblende-rich gabbro. When at Stop 2, look back for a good view of this.

15.4 **Stop 2. Cuyamaca-Laguna Mountain Shear Zone.**
1.6 [32.82713°N, 116.49575°W] [elevation 4,425 feet] Large parking area. Pull off to the left (west). The large rock is called "Graffiti Rock" by the locals.

This wide parking area affords a view of the town of Pine Valley and of the western zone of the Peninsular Ranges Batholith. The broad, conical edifice to the west is Viejas Mountain, an undated gabbro complex that sits above the Viejas Indian Reservation. The granitic rocks below the view area and off to the west belong to the 118 million year old Pine Valley Pluton, the only western zone rock with alkali feldspar phenocrysts and no magnetite.

The western zone of the batholith contains scattered exposures of deformed Jurassic granitic rocks that have been intruded by a variety of Cretaceous plutons (Figure 2). This belt of deformed rocks, the Cuyamaca-Laguna Mountain Shear Zone, consists of relatively light-colored, gneissic igneous rocks that were intruded into the sedimentary precursors of the Julian Schist which then lay at the western edge of the North American continent. These gneissic rocks are typically biotite-bearing granodiorite and tonalite with uranium-lead zircon ages that typically range between 170 and 161 million years ago, (Todd, 2004). In addition, the rocks also contain a small population of zircons with an average age near 1500 million years. Such a mixture of ages could come about only if the rising magmas intruded through the leading edge of the North American continent and incorporated fragments of those ancient rocks. In effect, the Jurassic gneisses represent a magmatic arc that collided with the western edge of the North American Craton approximately 160 million years ago.

Two separate uranium-lead ages from multiple zircon fractions were determined (Thompson and Girty, 1994) from the buff-colored Jurassic gneiss in a road cut further up the highway. The first gave a discordant age of 156 (±12) million years whereas the second produced a similarly discordant age of 256 (± 39) million years. This disparity and error margin illustrates the difficulty

Figure 4. Road Cut in Jurassic Gneiss along Sunrise Highway.

in accurately dating rocks with multiple zircon populations.

Structurally, the granitic rocks of the Cuyamaca-Laguna Mountain Shear Zone display a strong foliation that consistently strikes more or less northwestward and dips steeply to the northeast. Detailed studies indicate that the foliation was created during two distinct periods. The oldest is pervasive and compressional in nature with an east-over-west pattern. For Jurassic gneisses closest to the eastern side of the Cuyamaca-Laguna Mountain Shear Zone, however, the opposite is true. Here, a younger, east-side-down fabric is superimposed on the older foliation. Careful field work by Thompson and Girty has limited the two stages of fault movement to between 118 and 105 million years ago for the older compressional event and between 105 and 94 million years ago for the younger extensional phase. The significance of these two periods will be considered ahead.

The large elongate feature rising from the flat parking area was preserved during the road building process because it was thought to be an exposure of a fault surface. In fact, it is a granitic dike composed mostly of quartz and feldspar that shows little evidence of being caught up in any fault movement. Instead, the dike contains small, scattered pieces of gneissic rock and appears to have intruded into the surrounding gneisses.

The road continues through Jurassic granitic gneisses for the next several miles. A portion of the 118 million year old Pine Valley Monzogranite, which underlies the area just west of the highway, extends into and has been deformed by the Cuyamaca-Laguna Mountain Shear Zone. This relationship establishes the upper age limit for movement on the fault system.

18.2 **Stop 3. Chain-Up Area.** [32.85487°N, 116.47583°W]
4.4 [elevation 5,140 feet] Pull off to the right (east). The southern side of Crouch Meadow marks the southern contact between the 105 million year old Cibbett Flat Pluton of the western Peninsular Ranges Batholith and the Jurassic gneisses within the Cuyamaca-Laguna Mountain Shear Zone. The meadows here and inside the eastern loop of Sunrise Highway (Figure 2) are a curious topographic feature.

Figure 5. A Portion of Laguna Meadows Showing Scattered Outcrops of Tonalite.

18.8 On the left along a dirt road an example of spheroidal weathering can be seen but an excellent classic example will be seen further up the road at 19.7 miles.

19.1 Forest Service Information Station on Boardwalk (right side)

19.2 Pleasant hikes to the north and east along trails from the Meadows Information Kiosk on the left take you through Laguna Meadows and out to a series of small lakes and ponds. The entire area lies within the Cibbett Flat Pluton and piles of rounded tonalite boulders dot the landscape. Why the meadows persist within the forested area and what controls their location is not well documented. Nonetheless, they provide an idyllic setting for wildlife and day hikers alike.

19.7 **Stop 4. Turn out on south (right) side of Sunrise Highway.**
5.9 [32.85774°N, 116.45760°W] [elevation 5,588 feet] The outcrop across the road is part of the Cibbett Flat Pluton that has a reported zircon uranium-lead age of 105 million years. This pluton is large for a western zone body and covers nearly 40 square miles (100 km^2) (Figure 2). It consists of two or possibly three different and recognizable textural types of tonalite. The

Figure 6. Road Cut in Tonalitic Grus at Stop 4.

most common type along Sunrise Highway consists of variable amounts of pyroxene, amphibole, and biotite along with abundant plagioclase and quartz. The biotite has formed ¾-1¼ inch (2-3 cm) oikocrysts. Oikocrysts are single crystals that formed from small blobs of remnant magma late in its cooling history. With sunlight cascading down on the outcrop, the oikocrysts are easily visible from light reflected off their cleavage surfaces. Careful examination of these oikocrysts reveals that the biotite encloses tiny crystals of quartz, feldspar, and pyroxene indicating that it formed last in the crystallization history of the tonalite magma.

The outcrop is a classic example of spheroidal weathering. Weakly acidic groundwater percolating through a closely spaced, three-dimensional fracture system chemically attacked the edges and corners of the fracture blocks first. This resulted in a rounding of these outside parts of the blocks leaving a sphere-shaped, relatively unweathered core surrounded by loose material that once constituted solid tonalite. Close examination of this loose material reveals that it still has the same mineralogy as the fresh rocks but the grains have been sufficiently disaggregated and now are almost as loose as beach sand.

Figure 7. View at Mile Post 20. Photo by Dave Bloom.

20.0 Turnout on right. The southwestward view here is geographically confusing because of the twists and turns of Sunrise Highway. The rolling terrane below is underlain by tonalitic rocks in the Cibbett Flat Pluton. The community of Pine Valley is just out of sight below and a portion of Interstate Highway 8 heading back west towards San Diego can be glimpsed between the trees. To the right (northwest), the familiar profile of Cuyamaca Peak and its neighbor Middle Peak, both underlain by gabbro, are visible. The light-colored rocks in front of Middle Peak are part of the Stonewall Peak Granodiorite (see Stop 11). The dark peak on the western horizon (Figure 7) is Viejas Mountain, also underlain by gabbro. To its right, the light-colored ridge is El Capitan (which is designated as El Cajon Mountain on some maps), another intermediate composition granitic rock with an unusual smattering of disseminated molybdenum sulfides once targeted by the mining community.

21.0 Turnout on right. View southward into Mexico across the Cibbett Flat Pluton.

21.9 Morris Ranch Road. The San Diego State University Observatory is about 1/4 mile east.

23.5 Mt Laguna – here you will find the Laguna Mountain Lodge and Store/Trading Post and the Laguna Mountain Visitor Center.

23.7 Entrance to the Desert View Picnic Grounds.

Figure 8. View Southeastward from the Radar Station. Photo by Dave Bloom.

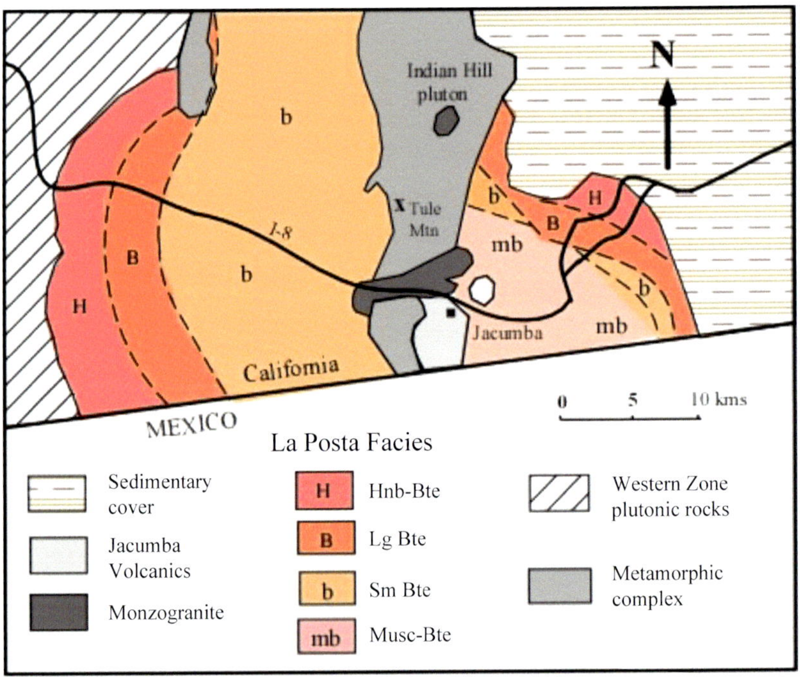

Figure 9. Generalized Geologic Map of the La Posta Pluton.

23.8 *10.0* **Stop 5. Right Turn Toward the Radar Station**. [32.87084°N, 116.41577°W] [elevation 5,969 feet] Continue ahead for 0.3 miles to an overlook at the end of the public portion of the road. [N 32.87264, W 116.41262] [elevation 6,052 feet] The view from this point extends eastward through the eastern zone of the Peninsular Ranges Batholith and beyond to the Salton Sea. On clear days, the Algodones Sand Dunes near Yuma are visible.

The darker, brush-covered ridge to the right (southeast) is the easternmost part of the Cibbett Flat Pluton (Figure 8). The rocky saddle at the end of that ridge is the northern end of the 94 million year old La Posta Pluton with the saddle marking the contact between the two zones. The La Posta Pluton (Figure 9) is the largest igneous body in this region and covers approximately 500 square miles (1400 km^2). It is concentrically zoned from a hornblende- and biotite-bearing tonalite margin through two distinctive biotite-rich zones into a muscovite-biotite monzogranite. Contacts between the zones are gradational suggesting that this large igneous mass was emplaced into a single magma chamber where it cooled from the margins inward. In doing so, the magma underwent a process called crystallization differentiation resulting in the inward variation in rock type.

The jagged ridge that extends eastward is the Sawtooth Range with Whale Peak and part of the Vallecito Mountains off in the distance. Both mountain ranges are part of the eastern zone of the batholith but Whale Peak lies on the eastern side of the Elsinore Fault and may be either a fault-displaced portion of the La Posta Pluton or a separate eastern-zone intrusion.

Recent paleomagnetic studies (Symons, et al, 2008) have suggested that the western zone of the Peninsular Ranges Batholith is an exotic terrane that drifted as much as 250 miles (400 km) to the northwest prior to its collision with the Jurassic arc (Figure 10). As the magmatic arc migrated towards North America along coast-parallel faults, it grew in size. Its volcanoes continued to erupt and slowly crystallized their root zones into the plutonic rocks we see at the surface today. During this collision, the western zone was sliced into discrete blocks along north-northwest-trending faults that were rotated, east-side up, by as much as ten degrees to produce the long-recognized north-northwesterly structural trend of the western zone.

Eastern Zone

The eastern zone of the Peninsular Ranges Batholith is structurally and lithologically distinct from the western zone. It is composed of very large (100's of sq. km.), concentrically zoned granitic plutons with little or no internal deformation. Eastern San Diego County, for example, is dominated by the large La Posta Pluton (Figure 9) that has an exposed area in excess of 500 square miles (1400 square km.). It grades inwards from a hornblende-bearing tonalite margin through several distinctive biotite-bearing granodiorites into a core zone of biotite-muscovite monzogranite.

Western zone rocks all contain magnetite and ilmenite as trace minerals whereas the eastern zone only has ilmenite. This seemingly minor change allows for the easy recognition of the two zones through regional

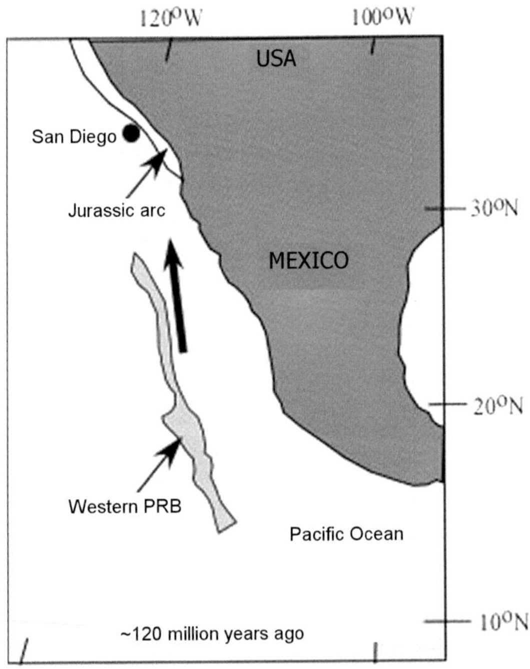

Figure 10. Approximate Position of the Proto-Western Zone as It Begins Its Northward Migration towards Southern California.

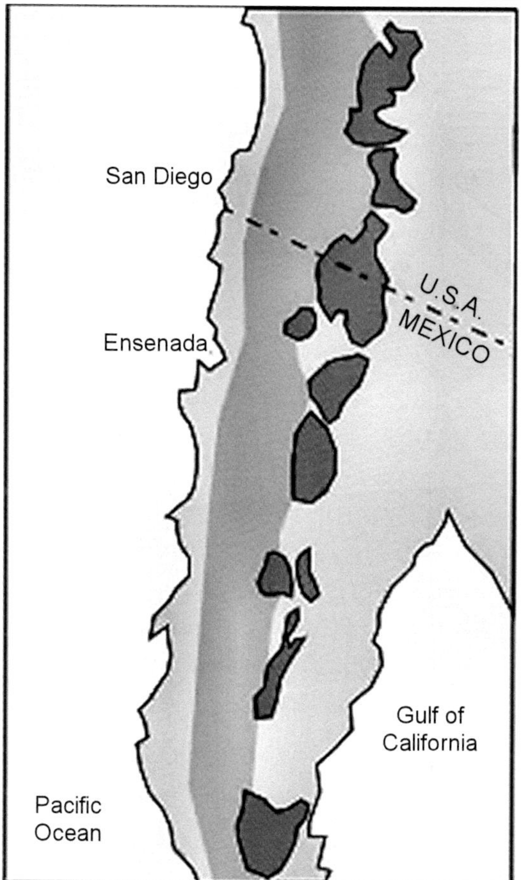

Figure 11. Distribution of 90-100 Million Year Old Granitic Plutons (dark) of the Eastern Zone of the Peninsular Ranges Batholith. Medium gray approximates the western zone of the batholith.

aeromagnetic patterns. The paucity of magnetite has also stymied paleomagnetic studies within the eastern zone. However, recently published studies indicate that the La Posta Pluton has essentially crystallized in place and thus contrasts with the inferred movement of the western zone.

Emplacement ages for these zoned plutons all along the strike length of the batholith (Figure 11) fall into a fairly tight age frame between 93 and 98 million years old and are distinctly younger than emplacement ages for the western zone.

Return to Sunrise Highway.

26.5 **Stop 6. View Platform Opposite the Al Bahr Shrine Camp.**
12.7 [32.89616°N, 116.44881°W] [elevation 5,472 feet] The view to the northeast from this point is directed at Whale Peak, part of the eastern zone of the Peninsular Ranges Batholith, and centers down Storm Canyon. The rocks on either side of the canyon are Jurassic gneisses of the Cuyamaca-Laguna Mountain Shear Zone. Large, white dikes cut the Jurassic gneisses along the southern side of the canyon. These dikes are granitic in composition and contain both biotite and muscovite. They commonly show a foliation parallel to that in the Jurassic gneisses indicating that they may be part of the Jurassic magmatic event. Some, however, crosscut the foliation and may be related to the granitic pegmatites that occur throughout the medial portion of the batholith. Near Pala and Mesa Grande, pegmatitic dikes contain a variety of gem minerals and collector specimens that include tourmaline, spodumene, lepidolite, and beryl. San Diego County is world-renowned for its production of these minerals and visitors to natural history museums and mineral shops everywhere can always find local specimens on display.

The trail that can be seen below the boardwalk is the Pacific Crest Trail which is a designated national scenic trail that stretches 2,650 miles from Mexico to Canada. Every April there is what is called Zero Day at Lake Morena where hikers will start their trek of approximately 5 months duration.

Just down the road is Foster Lodge, where two cabins are available for overnight stays for a nominal fee if you are a Sierra Club member.

The Al Bahr Shrine Camp was one of the first places developed after roads came to Mount Laguna in 1918. The Al Bahr Temple originally leased a 20-acre tract of land from the forestry service in 1921. The Mount Laguna Shrine Club was chartered in 1963. The camp has grown over the years and now has cabins, other facilities and a mess hall.

27.4 **Stop 7. Penny Pines Parking Area and Trailhead.** [32.90570° N, 116.45738° W] [elevation 5,450 feet] This stop consists of a 1.5 mile hike out and back along the Pacific Crest Trail to catch some of the best and least seen views in the county. The trail leaves from the south end of the parking area and winds its way eastward toward the northern rim of Storm Canyon (Figure 12). It merges with the Pacific Crest Trail at approximately two hundred feet from the parking area and continues around the eastern side of a broad hill towards Garnet Peak. This hill is underlain by a modest-sized gabbroic body with exposures both along and above the trail.

13.6

After winding about 500 feet through a burned area from a fire in 2003, the trail emerges along the northern side of Storm Canyon. The first outcrops encountered are reddish to buff-colored Jurassic gneisses with small ellipsoidal inclusions of Julian Schist. At this point, the trail is covered with weathered cobble-sized pieces of the gneiss. About 100 feet further ahead, the character of the trail changes abruptly as it crosses the contact between the gneiss and the more weathering-resistant Cretaceous

Figure 12. View Southeastward across Storm Canyon. Photo by Philip T. Farquharson.

Figure 13. Outcrop of Jurassic Gneiss at End of Side Trail for Stop 7. [32.91154° N, 116.45275° W] [elevation 5,544 ft.] Photo by Philip T. Farquharson.

gabbroic rocks. From this point onward for about another quarter mile, the trail is cut through amphibole-rich gabbroic rocks. The first large loose boulders encountered are characterized by ½ to 1 inch (1-2 centimeter) amphibole oikocrysts, large single crystals that formed late in the cooling history of the melt. They contain numerous inclusions of early formed white, rectangular plagioclase. A short secondary trail another 100 feet ahead to the right (southeast) leads to a rocky point with sweeping views of the canyon and the desert below (Figure 13). Some of the loose boulders here also contain reddish olivine and pyroxene (along with the whitish plagioclase) as inclusions within the amphibole. Their reddish color is due to surface weathering of these iron-bearing minerals.

The main trail turns more northerly at this point and continues through the gabbro towards a saddle about ¼ mile ahead. The rocks found along the trail are still part of the gabbro body on the hill above the trail. Grain sizes vary considerably in these rocks from sub-millimeter black amphibole-rich rocks to whitish, coarser-grained plagioclase-rich phases. This variation is due to internal processes within the magma chamber as the melts cooled and crystallized.

At the saddle, the character of the trail changes again as it again crosses the contact between the gabbro and the Jurassic gneiss. Follow the side trail a few hundred feet to the right towards a large outcrop of Jurassic gneiss and associated granitic rocks (Figure 13). From this point, the desert views are magnificent. Whale Peak, in the eastern zone of the batholith, is the large mountain ahead (east). The Elsinore Fault Zone lies at its base. The Vallecitos Mountains are visible over the southern edge of Whale Peak and lead southward into the Fish Creek and Coyote mountain ranges. A view well worth the short hike.

At this point, biotite- and muscovite-bearing granitic rocks appear to be intimately associated with the Jurassic gneiss. Close examination of the gneiss shows small granitic stringers and thin bands of nearly pure biotite parallel to the gneissosity. This association suggests that parts of the gneiss have undergone a small degree of partial melting to produce a composite rock known as migmatite.

Return along the trail to Sunrise Highway. Continue north another 2.9 miles to Kwaaymii Point Road. Turn right and proceed another ¼ mile to the overlook and occasional jumping-off point for hang gliders and parasailers.

30.3 **Stop 8. Kwaaymii Point and Cuyamaca-Laguna Mountain**
16.5 **Shear Zone.** [32.93445° N; 116.48259° W] [elevation 5,483 feet]

The view from the parking area is one of the most spectacular in the county. The green area below is Agua Caliente Springs and the large prominent mountain beyond the town site (Whale Peak) is part of the eastern zone of the batholith. The Elsinore fault zone runs along the mountain's southwest facing slopes as it heads back towards Banner Grade and the town of Julian. The distant peaks on the horizon are the Santa Rosa Mountains, which are also part of the eastern zone of the batholith.

Figure 14. Rockfall across Remnant of Old Sunrise Highway.

One can take a short walk along the abandoned roadway, which is now part of the Pacific Crest Trail, to examine the nature of the deformed Jurassic granitic rocks that mark the Cuyamaca-Laguna Mountain Shear Zone. The roadway is what remains of the old portion of Sunrise Highway. It was repeatedly threatened by rock falls from the adjacent cliffs. The main road was shifted to its present course in the late 1970s but one rock fall remains across the road a few hundred yards south of the parking area (Figure 14).

The rocks consist of cream-colored feldspars, quartz, and biotite that have been strung out into a steep, east-dipping foliation that locally parallels the cliff face (Figure 15). Two periods of deformation have been recorded (Thompson and Girty, 1994). The oldest, bracketed between 118 and 105 million years ago, shows east over west microstructures indicating a compressional regime. The youngest, found only along the eastern margins of the shear zone and superimposed on the older rock fabric, is extensional with east-side down movement. Timing of this latest

Figure 15. Two-Foot Long Inclusion of Julian Schist in Jurassic Gneiss near Intersection of Sunrise Highway with Kwaaymii Point Road.

movement has been shown to be between 105 and 94 million years ago. The older deformation was likely caused by the collision of the western zone with the continent-fringing Jurassic arc whereas the younger is related to the emplacement of the 94 million year old La Posta Pluton. The contact between the two lies in the saddle, half way down the ridge. Return to Sunrise Highway, turn north.

31.1 *17.3* **Stop 9. Pegmatite dike**. [32.94495° N, 116.49427° W] [elevation 5,467 feet] Parking area on right (north) side of Sunrise Highway.

The road cut across from the east end of the parking area exposes weathered Jurassic gneiss cut by a sinuous pegmatitic dike (Figure 16). The foliation in the gneiss exhibits a roughly north-south strike that at this point is approximately perpendicular to the highway and dips steeply to the east. Numerous ellipsoidal pieces of Julian Schist have been caught up in the gneiss and are aligned with the foliation. Some appear to be dark schistose rocks whereas others are dark, fine grained quartzite.

Figure 16. Pegmatite Dike Cutting Weathered Jurassic Gneiss

The dike consists mainly of gray quartz and large, cream-colored blocky feldspar similar to what was seen at the first stop. It cuts across the foliation in the gneiss and, if it represents the same event as the dike at Stop 1, is also younger than the 118 million year old Pine Valley Pluton. A small, near vertical fault cuts the gneiss and slightly offsets the dike just before it disappears from the road cut (Figure 16).

On a clear day, the observatory domes at the summit of Palomar Mountain can be seen to the north. To the northeast, the western slopes of Vulcan Mountain drop down into Banner Grade and the Elsinore Fault which continues southward below Granite Mountain, the prominent topographic feature to the east. Granite Mountain is a "La Posta-type" pluton in the Eastern Zone of the batholith.

32.5 **Stop 10. Oriflamme Canyon Protomylonite**. [32.955878°
18.8 N, 116.50896° W] [elevation 5,158 feet] The outcrops at the southern end of a paved pullout are part of the Oriflamme Canyon Pluton. This unit and the Lucky 5 Pluton just ahead were studied by Melanie Briggs and other students at San Diego State University. Outcrops of deeply weathered Julian Schist across the road from the pullout represent the country rock into which both rock units were emplaced.

The Lucky 5 Pluton consists of a central core of undeformed biotite-hornblende granite/granodiorite that grades into a narrow margin of biotite-muscovite granite. Zircons from this pluton yielded an age of 130 million years old but older studies at Cal Tech reported an age of 115 million years old. The published geologic map (Figure 2) shows the same unit on either side of the highway. The San Diego State University project, however, mapped the unit east of Sunrise Highway as the Oroflamme Canyon Pluton, a biotite-rich granitic rock with a pronounced mylonitic fabric (Figure 17). It yielded a zircon age of 116 million years old. How these two units, with their disparity in age and fabric, are related and how they fit into the overall tectonic scheme is uncertain.

Figure 17. Outcrop of Oriflamme Canyon Mylonite at Stop 10. Note coarse grain size and pronounced eastward dipping foliation.

Figure 18. Contact between Lucky 5 Pluton and Julian Schist along Sunrise Highway. Photo by Philip T. Farquharson.

33.0 The contact between Julian Schist and the Lucky 5 Pluton is at the rocky ridge to the left (west). The steep bouldery hillside just ahead is also part of the Lucky 5 Pluton whose outcrops skirt the west side of the highway (Figure 18).

33.5 Steep road cuts in Julian Schist with amphibolite over the next mile (Figures 19 and 20).

Figure 19. Road Cut in Julian Schist.

This site is not listed as a trip stop because parking is limited at best and the shoulder very narrow. If you wish to examine this road cut, park ahead in one of two places where the shoulder is wide enough for a single car to pull off. Walk back very cautiously to the outcrops.

Figure 20. Road Cut in Julian Schist.

Julian Schist

The oldest known rocks within the region are scattered outcrops of a Triassic metasedimentary assemblage known collectively as the Julian Schist (Hudson, 1922). This unit exhibits a lateral variation in composition as one travels from the western to the eastern zone of the batholith. In the west, it consists mainly of biotite-bearing quartzite with lesser amounts of biotite schist and quartz-pebble conglomerate. Near the boundary between the western and eastern zones, more schistose rocks, some muscovite-bearing, appear and form a thick sequence with alternating layers of schist and quartzite. Still further to the east, near the San Diego-Imperial County line, the character of these rocks changes abruptly. The amount of schistose rocks has decreased, the quartzites contain little or no micaceous minerals, and marbles now form a significant portion of this older rock package. This eastward change in rock type is accompanied by an increase in the degree of metamorphism. In the west, metamorphic minerals such as andalusite and cordierite suggest metamorphic conditions near 600°C and 2.5 kilobars of pressure. In the eastern zone, however, minerals such as sillimanite, diopside, and wollastonite indicate temperatures of 640°C and total pressures between 3.5 to 4.0 kilobars.

Finally, scattered throughout all portions of the Julian Schist are isolated exposures of dark, fine-grained amphibolite, a metamorphic rock composed of hornblende and plagioclase. These rocks are the metamorphosed equivalent of basalt and represent lava flows, dikes, or sills within the Julian Schist. The chemistry of these metamorphosed basalts is distinctive and can be correlated to basaltic rocks erupted at mid-ocean spreading centers. Thus, it seems that the depositional basin for the sediments that eventually became the Julian Schist must, at some point, have endured a spreading or rifting event that created the intimate spatial association of sedimentary and basaltic igneous rocks. These basaltic rocks, now amphibolites, are the oldest known igneous rocks in our area and represent a tectonic setting quite different from and much in advance of that which produced the Peninsular Ranges Batholith.

Figure 21. Entering Cuyamaca Meadows at Mile 36.0. Photo by Philip T. Farquharson.

36.0 **Turnout to Pedro Fages Monument.** [32.99300° N, 116.54830°W] [elevation 4,762 feet] The next 1.6 miles of Sunrise Highway has been adopted by the San Diego State University Geology Alumni Group, who can bee seen every three months in their safety vests picking up roadside litter.

36.5 **Stop 11. Cuyamaca Meadows.** [32.99799° N, 116.54830° W]
27.7 [elevation 4,654 feet] Parking is very limited along this portion of Sunrise Highway but a few places allow for a brief stop along the shoulder. The large, darkish mountain (Cuyamaca Peak) to the left (west) is underlain by gabbro. This plutonic type can usually be recognized by its topographic prominence, paucity of outcrops, and dark reddish soils. The white bouldery outcrops are the monzogranites of Stonewall Peak. Neither of these two plutons has been dated using radiometric techniques. The meadow itself is underlain by a mixture of Jurassic gneisses and Julian Schist.

Figure 22. View across Cuyamaca Meadows towards North Peak (gabbro). Photo by Philip T. Farquharson.

37.6 End of Sunrise Highway. Right turn (north) on SR 79 toward
23.9 Julian and proceed 3.1 miles to final stop at Inspiration Point.

Figure 23. Inspiration Point. Panoramic photo by Philip T. Farquharson.

Stop 12. Vista Point (a.k.a. Inspiration Point). [33.05009° N;
27.0 116.56859° W] [elevation 4,656 feet]

The few weathered outcrops at this view point are granitic rocks of the western zone of the Peninsular Ranges Batholith. In the distance to our left (northwest) are the observatory domes housing the great telescopes at Palomar Mountain. Straight ahead (northeast), the prominent topographic feature is Vulcan Mountain, part of the eastern zone of the batholith. The Elsinore Fault, whose total right-lateral strike-slip movement appears to be approximately 1.2 miles and whose age is estimated at less than 2 million years old, follows the steep-sided valley immediately below. In this area, the Elsinore Fault passes through a metasedimentary roof pendant belonging to the Julian Schist. The gold deposits upon which Julian was founded all occur as auriferous quartz veins cutting the metamorphic rocks which underlie the hills below the view point. Fluid inclusion temperatures and isotopic data indicate that the gold deposits formed as a result of fluids released during the breakdown of hydrous minerals within the metamorphic assemblage.

Mining

Mining is a very risky proposition, subject to many and varied factors that can prevent a single mine or even a district from fulfilling its economic potential. The Julian district faced more than its share. A disputed boundary for the Rancho Cuyamaca land grant tied the independent miners up in court for several years and prevented them from directing their capital into improving the mines and milling operations around Julian. Although the miners eventually won their case in court, they were not able to recover from that financial blow, in part, because lending institutions in southern and even northern California refused to lend money to Southerners. The free-milling gold that characterized most of the district and made the extraction of the gold from the quartz ore so profitable turned into a sulfide ore at depth. The sulfide ore could only be processed with new, expensive smelters and money was in short supply. The mines fell into disrepair and many flooded as groundwater slowly filled the underground workings. Even the discovery of the district's richest lode deposit in 1871, the Golden Chariot mine, could not alter the inevitable and by the mid to late 1870s, the boom was over. There was a minor resurgence in 1888 with the discovery of the Gold King and Gold Queen mines just south of the main belt of mines in the district but the easily extracted ore soon ran out and the boom gave way once again to bust. Today, independent miners or small partnerships still struggle to recover the precious metal from a few mines in the district but the glory days are long over.

Branching off from the Elsinore Fault near the break in slope is Chariot Canyon and the Chariot Canyon Fault. This older feature appears to have been reactivated during movement on the Elsinore Fault and has caused some mechanical granulation within the gold-bearing quartz veins in the Chariot Canyon segment of the roof pendant. The Elsinore Fault also marks the local boundary between the western and eastern zones of the Peninsular Ranges Batholith. The building structures behind us sit on weathered olivine gabbro. Note the typically reddish soils along the road.

The broad valley to the east is the apt-named Earthquake Valley, which has been called Shelter Valley and is sometimes now called Paradise Valley by developers trying to sell land and homes further down the valley. Beyond are the Vallecitos Mountains and the Pinyon Mountains which overlie portions of a large detachment fault.

Summary and Tectonic Development

The composite nature of the Peninsular Ranges Batholith was created by a series of events that began with the formation of the Jurassic granitic rocks that now form the topographic backbone of the Laguna Mountains. 160 or 170 million years ago, a subduction zone created an island arc system and its associated plutonic roots at the western edge of the North American continent. These magmas intruded sedimentary rocks of at least Triassic age, the remnants of which are also found as isolated roof pendants within all of the Mesozoic portions of the batholith. Beginning at least 120 million years ago, a second arc developed well to the south. This system, what is now the western zone of the Peninsular Ranges Batholith, grew larger and moved northward until it collided with the Jurassic arc at about 110 million years ago to create the older compressional fabric in the Cuyamaca-Laguna Mountain shear zone. Subduction continued until about 105 million years ago at which point, the subduction plane shallowed and magmatism in the western zone ceased.

With a shallower angle of subduction, magmatism moved inland (eastward). Why and under what conditions it now produced such copious amounts of homogeneous melt along the strike-length of the batholith is unclear. The rise of the large La Posta plutons released their latent heat of crystallization into the metasedimentary country rocks leading to the breakdown of micaceous minerals. Partial melting within the metasedimentary assemblages under high P_{H_2O} conditions created the pegmatite dikes and other small bodies of muscovite-bearing granite. The release of excess fluids during this process may be responsible for the leaching of trace amounts of gold from ancient placers and their subsequent concentration at shallower levels in auriferous quartz veins. As the country rock rose, piston-like, in response to the invading La Posta magmas, the second extensional fabric was overprinted onto the original compressional fabric in the adjacent Jurassic rocks.

At some point after approximately 90 million years ago, the subduction plane again shallowed, causing eastern zone magmatism to end. Eastward-directed subduction at this new angle likely went on to develop the Laramide events of central Arizona and beyond.

Bibliography and Selected Readings

Abbott, P.L., 1999. The Rise and Fall of San Diego: 150 Million Years of History Recorded in Sedimentary Rocks. San Diego: Sunbelt Publications, 231 p.

Biggs, M., 2007, The Lucky 5 Pluton and Oriflamme Canyon Plutons: Key temporal markers in the structural and tectonic evolution of the Peninsular Ranges Batholith, SE California, abs., S.D.S.U. senior field report.

Bloom, D.M., P.T. Farquharson, and C.L. Ziegler, editors, 2006. Geology and History of Southeastern San Diego County, California, San Diego Association of Geologists Field Trip Guidebook for 2005 and 2006, 208 p.

Fetzer, L., 2002. A Good Camp: Gold Mines of Julian and the Cuyamacas. San Diego: Sunbelt Publications, 106 p.

Houser, C.E., and M.L. Murbach, editors, 2010. Geology and Lore of Northern Anza-Borrego Desert Region. San Diego Association of Geologists and South Coast Geological Society Volume 37, 198 p.

Hudson, F.S., 1922, Geology of the Cuyamaca region of California: Calif. Univ., Dept. of Geol. Sci. Pub., v.13, pp. 175-252.

Murbach, M.L., and M.W. Hart, editors, 2003. Geology of the Elsinore Fault Zone, San Diego Region. San Diego Association of Geologists and South Coast Geological Society Volume 31, 212 p.

Symons, D.T.A., Smith, T.E., Kawasaki, K., and Walawender, M.J., 2009, Paleomagnetism of the mid-Cretaceous gem-bearing pegmatite dikes of San Diego County, California, USA; Canadian Journal of Earth Science, v. 46, No. 9, pp. 675-687.

Symons, D. T. A., Walawender, M. J., Smith, T.E., Molnar, S.E., Harris, M. J., and Blackburn, W.H., 2003, Paleomagnetism and geobarometry of the La Posta Pluton, California: in Johnson, S.E,. Paterson, S.R., Fletcher, J.M., Girty, G.H., Kimbrough, D. K., and Martin-Barajas, A., eds., Tectonic evolution of northwestern Mexico and southwestern U.S.A., Geological Society of America Special Paper 374, pp. 135-155.

Symons, D.T.A., Smith, T.E., and Blackburn, W.H., 2008, Constraining the travels of a suspect terrane: Paleomagnetism and geobarometry of two Early Cretaceous igneous complexes in the Peninsular Ranges Batholith, California; Geological Society of America Bulletin September 2008, v. 120, pp. 1159-1170.

Streckeisen, A. L., 1974. Classification and Nomenclature of Plutonic Rocks. Recommendations of the IUGS Subcommission on the Systematics of Igneous Rocks. Geologische Rundschau. Internationale Zeitschrift für Geologie. Stuttgart. Vol.63, pp. 773-785.

Thompson, C. N. and Girty, G. H., 1994, Early Cretaceous intra-arc ductile strain in Triassic-Jurassic and Cretaceous continental margin arc rocks, Peninsular Ranges, California: Tectonics, v. 13, pp.. 1108-1119

Todd, Victoria, R., 2004, Preliminary Geologic map of the El Cajon Quadrangle, Southern California; U.S.G.S. Open File Report 2004-1361.

Walawender, Michael J., 2000. The Peninsular Ranges: A Geological Guide to San Diego's Back Country. Dubuque: Kendall/Hunt Publishing Co., 124 p.

Walawender, M. J. , Gastil, R. G., Clinkenbeard, J. P., McCormick, W. V., Eastman, B. G., Wernicke, R.,S., Wardlaw, M. S., Gunn, S. H., and Smith, B. M., 1990, Origin and evolution of the zoned La Posta-type plutons, eastern Peninsular Ranges Batholith, southern and Baja California, in Anderson, J. L., ed., The nature and origin of Cordilleran magmatism: Geological Society of America Memoir 174, p. 1-18.

Weber, F.H., Jr., 1963. Miners and Mineral Resources of San Diego County, California. California Division of Mines and Geology County Report 3, 309 p.

Winter, John D., 2001, An Introduction to Igneous and Metamorphic Petrology: Prentice Hall, Upper Saddle River, New Jersey 07458, 697 p.

Glossary

Amphibolite: A medium- to high-grade metamorphic rock consisting primarily of plagioclase and amphibole.

Basalt: A dark-colored, fine-grained volcanic rock with less than 52 % SiO_2.

Batholith: An exposed, irregular mass of igneous rock greater than 100 sq.km. in total area and consisting of individual plutons.

Calc-alkaline: A chemical classification of a suite of igneous rocks such that the total alkali content (Na_2O + K_2O) and calcium content are equal between 52 and 56% SiO_2.

Compression(al): A stress regime in which the rocks are being squeezed together.

Craton: A mass of older igneous and metamorphic rocks that forms the nucleus of all continents, e.g. the North American craton.

Crystallization differentiation: An internal process in a magma chamber that separates early formed crystals from the parent melt, thus changing the composition of the remaining (daughter) melt.

Detachment fault: A shallow-dipping, normal fault of regional extent.

Dike: A tabular mass of intrusive igneous rock that cuts across older rock structures.

Discordant: Any geologic feature or structure that cuts across preexisting rock structures, e.g., a dike or fault.

Extension(al): A stress regime in which rocks are being pulled apart.

Fabric: An alignment of minerals or features in a rock generally due to recrystallization during metamorphism.

Felsic: A rock dominated by light-colored silicate minerals, especially quartz and feldspars.

Fluid inclusion: A small amount of entrapped fluid, e.g. water, in a crystal.

Foliation: A term for the alignment of mineral grains into a planar array.

Gabbro: A dark igneous rock consisting of olivine, pyroxene, amphibole and plagioclase. Gabbro is the coarse-grained (slowly cooled) equivalent of basalt.

Gneiss: A metamorphic rock characterized by an alignment of mineral grains into a linear array.

Granite: A light-colored igneous rock consisting of quartz, alkali feldspar and plagioclase feldspar.

Grus: fragmental products of in-situ granular disintegration of granite and granitic rocks. Also called "d.g.", for decomposed granite.

Intrusive: igneous rock that is the result of cooling and crystallization below the surface of the Earth.

Island arc: An arcuate belt of volcanic islands that marks the subduction of one lithospheric plate beneath another. Same as magmatic arc, volcanic arc.

Kilobar: A measure of lithostatic pressure due to the weight of the overlying rock column. One kilobar equals approximately 3.5 km depth in the earth.

Marble: A metamorphic rock consisting almost entirely of calcite.

Metamorphism: The process by which older rocks are recrystallized through heat and pressure into a new mineral assemblage or alignment.

Metasediment(ary): Any metamorphosed sedimentary rock.

Mid-ocean spreading centers: A linear belt along the sea floor where two lithospheric plates are being split apart and new material in the form of basalt magma is rising.

Migmatite: A "mixed rock" formed by partial melting of preexisting metamorphic rocks. It contains both the solidified molten material and the refractory or unmelted residue of the original rock.

Monzogranite: A variety of granite in which alkali feldspar and plagioclase are in approximately equal proportions.

Mylonitic: A textural term referring to an extreme alignment of mineral grains, generally as a result of movement within a fault plane.

Paleomagnetism: A branch of the Geological Sciences that examines the alignment of magnetic minerals in a rock due to growth in the earth's magnetic field. The data can be interpreted in terms of the movement of lithospheric plates with time.

Pegmatite: A very coarse-grained igneous rock, commonly granite.

Pluton: A body of intrusive igneous rocks generally less than 100 sq. km. in area.

Quartzite: A metamorphic rock consisting mainly of quartz and feldspar.

Radiometric age: A calculated absolute age of rocks and minerals using various radiogenic isotopes, e.g., uranium and lead.

Right-lateral: Movement pattern on a strike-slip fault in which the opposite side appears to have moved to the right.

Roof pendant: A large package of metamorphic rock completely encased in a younger igneous rock.

Schist: A metamorphic rock characterized by an alignment of mineral grains into a planar array.

Strike-slip: A type of fault in which the fault plane is vertical and motion is horizontal. See also "right lateral".

Subduction zone: A linear belt in which one lithospheric plate is pushed (subducted) beneath another. See also "island arc".

Tonalite: A plutonic rock containing quartz and plagioclase feldspar, but very little alkali feldspar.

Mineralogy Index

Alkali feldspar: $(K,Na)AlSi_3O_8$. The variety of feldspar with potassium and sodium It has three common forms with the same chemical composition, sanidine, orthoclase, and microcline. Sanidine is the high-temperature form that occurs only where quenched in volcanic rocks. Orthoclase and microcline occur in successively lower temperature plutonic rocks.

Andalusite: Al_2SiO_5. A metamorphic mineral formed by the reconstitution of clay minerals at low pressures and temperatures.

Beryl: $Be_3Al_2Si_6O_{18}$. A rare igneous mineral usually found in pegmatites. Its colored varieties include aquamarine (pale blue-green), emerald (dark green), heliodor (yellow), and morganite (pink).

Biotite: $K(Mg,Fe)_3(AlSi_3O_{10})(OH)_2$. The dark colored, iron- and magnesium-bearing member of the mica family.

Cassiterite: SnO_2. An ore of tin.

Cordierite: $(Fe,Mg)_2Al_4Si_5O_8$. A metamorphic mineral that is diagnostic of low temperatures and pressures.

Diopside: $Ca(Fe,Mg)Si_2O_6$. A member of the pyroxene family, it is a common igneous and metamorphic mineral. As a metamorphic mineral, it forms at moderate to high pressures and temperatures.

Hornblende: $NaCa_2(Mg,Fe,Al)_5(Al,Si)_8O_{22}(OH)_2$. The most common member of the amphibole family of minerals. It is found in both igneous and metamorphic rocks.

Ilmenite: $FeTiO_3$. An oxide of iron and titanium that forms under relatively reducing chemical conditions. See also magnetite.

Lepidolite: A member of the mica family that is commonly characterized by a lilac to dark purple color.

Magnetite: Fe_3O_4. An oxide of iron that forms under relatively oxidizing chemical conditions. See also ilmenite.

Muscovite: $KAl_2(Al,Si_3)O_{10}$ $(OH)_2$. The light colored member of the mica family.

Olivine: (Mg,Fe)SiO_4. A common igneous mineral in peridotite and basalt

Plagioclase: A member of the feldspar family consisting of a compositionally continuous series between albite ($NaAlSi_3O_8$) and anorthite ($CaAl_2Si_2O_8$)

Pyroxene: A family of silicate minerals with a general formula $Y_2(Si,Al)_2O_6$ where Y = Ca, Na, Mg, Fe, Mn). Pyroxenes occur in both igneous and metamorphic rocks. See also diopside.

Quartz: SiO_2. The second commonest mineral in the earth's crust (after feldspar).

Sillimanite: Al_2SiO_5. A metamorphic mineral formed by the recrystallization of clay minerals at moderate to high pressures and temperatures.

Spodumene: $LiAlSi_2O_6$. A rare mineral found in pegmatites. The gemmy green variety is hiddenite, the pink to violet variety is kunzite.

Tourmaline: $Na(Mg,Fe)_3Al_6(BO_3)_3(Si_6O_{18})(OH)_4$. A compositionally variable mineral found as an accessory in pegmatites and some hydrothermally altered rocks.

Wollastonite: $CaSiO_3$. A metamorphic mineral commonly found in marble that forms under moderate temperatures and pressures.

Zircon: $ZrSiO_4$. A common trace mineral in igneous rocks. It contains very small amounts of uranium (U) that can be utilized for U-Pb radiometric dating techniques.

About the Author

Michael Walawender was born in Auburn, New York. He earned a B.S. degree in Geology from Syracuse University, an M.S. degree in Geology from the South Dakota School of Mines and Technology, and a Ph.D. in Mineralogy and Petrology from The Pennsylvania State University. He began his career in the mining industry looking for metallic ore deposits in Labrador, Minnesota, Wisconsin, and Colorado. He then went on to spend 32 years as a professor of Geological Sciences at San Diego State University in San Diego, California. From 1989 through 1994, he served as Head of the Department of Geological Sciences at San Diego State. He became Professor Emeritus in 2004. He published scientific articles in a wide variety of national and international journals and, in 2000, wrote the book *The Peninsular Ranges: A Geological Guide to San Diego's Back Country.*

Dr. Walawender had a great love of teaching and following the progress of his students. He supervised nearly 100 Masters and Bachelors theses and his students have given many testimonials to the positive influence that he had on their lives. He led numerous field trips for both students and professional geologists throughout the Peninsular Ranges of Southern California. Though the author loved the beauty of the wildflowers of San Diego County, upon retiring from teaching, the author moved to central New York State to be close to family and enjoy the beauty of the Finger Lakes Region with his wife, Sonja, of 43 years. They have one son, Joshua.

(inside back cover, opposite) Color version of Figure 2, Field Trip Stops and Geologic Map of a Portion of the Cuyamaca-Laguna Mountain Shear Zone (see text, after Todd, 2004). Adapted by David M. Bloom.

Additional reference: Northern portion of map adapted from Plate 1: Geologic map of the Santa Ysabel quadrangle, California, in Merriam, Richard, 1958, Geology and mineral resources of Santa Ysabel quadrangle, San Diego County, California. California Division of Mines and Geology Bulletin 177.